Energy 130

大耳朵
Big Ears

Gunter Pauli

[比] 冈特·鲍利　著

[哥伦] 凯瑟琳娜·巴赫　绘

贾龙慧子　译

上海远东出版社

丛书编委会

主　任：田成川

副主任：闫世东　林　玉

委　员：李原原　祝真旭　曾红鹰　靳增江　史国鹏
　　　　梁雅丽　孟小红　郑循如　陈　卫　任泽林
　　　　薛　梅　朱智翔　柳志清　冯　缨　齐晓江
　　　　朱习文　毕春萍　彭　勇

特别感谢以下热心人士对童书工作的支持：

匡志强　宋小华　解　东　厉　云　李　婧　庞英元
李　阳　梁婧婧　刘　丹　冯家宝　熊彩虹　罗淑怡
旷　婉　王靖雯　廖清州　王怡然　王　征　邵　杰
陈强林　陈　果　罗　佳　闫　艳　谢　露　张修博
陈梦竹　刘　灿　李　丹　郭　雯　戴　虹

目录

Contents

小非洲象觉得很伤心，因为她的好些动物朋友都在嘲笑她。大家都拿她的大耳朵开玩笑。于是她用颤抖的声音问妈妈："我知道我们的耳朵特别大，但难道因为耳朵大，我们就不漂亮了吗?"

An African elephant calf is sad because some of her animal friends are teasing her. They are making fun of her large ears. In a trembling voice she asks her mum: "I know we have very big ears, but does that mean we are not beautiful?"

小非洲象觉得很伤心……

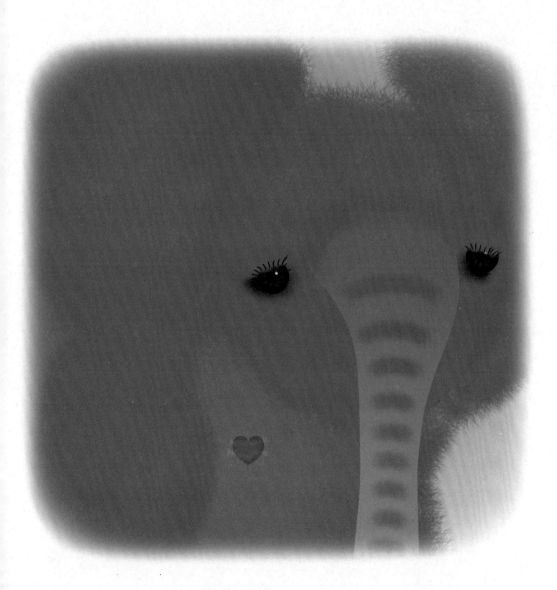

An African elephant calf is sad ...

你爸爸就很爱我的大耳朵……

your father loves my big ears ...

"情人眼里出西施。"象妈妈回答道，"你爸爸就很爱我的大耳朵，这就是他娶我的原因！"

　　"他可能是没有别的选择。"小象继续道，"我记得你告诉过我，我们非洲象是世界上耳朵最大的象。其他动物都比不过我们，甚至长耳大野兔也不行。"

"Beauty is in the eyes of the beholder," replies her mum. "Your father loves my big ears, that is why he married me!"

"He probably had no other choice," the calf replies and then continues, "I remember you told me that we African elephants have the biggest ears of all elephants in the world. And compared to those of other animals, no one beats us, not even the jackrabbit."

"你说得没错，我们耳朵的大小差不多是我们身体的六分之一。"

"六分之一！妈妈，那真是太大了！即使在五千米以外，我肯定也能把你和爸爸的声音听得清清楚楚。"

"是什么让你觉得我们的耳朵只是用来听声音的？"

"You are right, our ears probably are one-sixth of our body size."

"One-sixth! Mum, that's huge! I am sure I can hear you or Dad perfectly well, even from five kilometres away."

"What makes you think that we have ears only to listen?"

......我们身体的六分之一。

... one-sixth of our body size.

我的耳朵能当遮阳伞用。

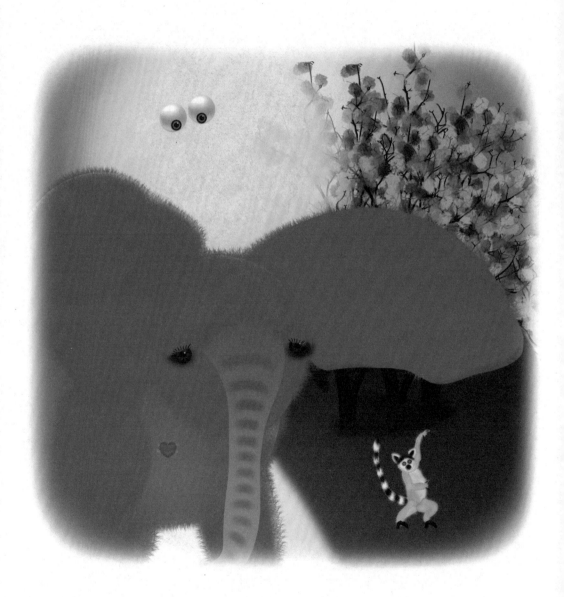

My ears could serve as a parasol.

"咦，耳朵不就是让声音进来并传向我们大脑的地方吗？"

　　"是这样的。但是当我们走在热带草原上时，耳朵还能为我们做些什么呢？"

　　"我的朋友们嘲笑我，说我的耳朵大到能当遮阳伞用。"

"Well, are ears not the place where sound enters and is funnelled to our brain?"

"Yes, but when we are walking across the savanna, what else do our ears do for us?"

"My friends tease me and tell me that my ears are so big that they could serve as a parasol."

"你的耳朵比任何遮阳伞都有用。遮阳伞只能提供阴凉，但你扇动耳朵可以让身体降温。拥有巨大的耳朵让我们即使在最炎热的时候，也能在长距离走动的同时保持凉爽。"

"一路走到水潭边，然后尽情地戏水！"

"Your ears do more than any parasol could. A parasol can only provide shade, but your flapping ears cool your body down. Having big ears allows us to remain cool and walk long distances – even on the hottest day."

"All the way to the waterhole, for a good splash!"

……你扇动耳朵可以让身体降温。

... you flapping ears cool your body down.

……厚度却只有一毫米。

... only one millimetre thin.

"说对了，我们大象的确喜欢好好冲个凉。当我们的耳朵湿了，热量散发出去的速度会更快，立刻就能让我们凉快起来。"

　　"但是我们的耳朵是灰色的，我听说深色更容易吸收热量。"

　　"嗯，你说得对。但是，尽管覆盖我们身体的皮肤都很厚，覆盖耳朵的皮肤厚度却只有一毫米。"

"Yes, we elephants do enjoy taking a long bath. When our ears are wet, the heat radiates out even faster, cooling us right down."

"But we have grey ears, and I have learned that dark colours absorb heat."

"Well, that is true. But while we have thick skin over the rest of our body, the skin over our ears is only one millimetre thin."

"那么薄？那以后我从荆棘丛里摘水果和树叶的时候可真要小心点，别伤到自己的耳朵。"

"确实如此。你的血液从耳朵里密布的静脉和动脉血管中流过，释放热量并让你身体内部的温度降下来。而且还能防止你身体里的水分流失。"

"That thin? I should then be careful not to cut my ears when trying to get to the fruit and leaves from thorn bushes."
"Indeed. Your blood runs through thousands of veins and arteries in your ears, releasing heat and cooling down the inside of your body. And that keeps you from losing water from your body."

……密布的静脉和动脉血管……

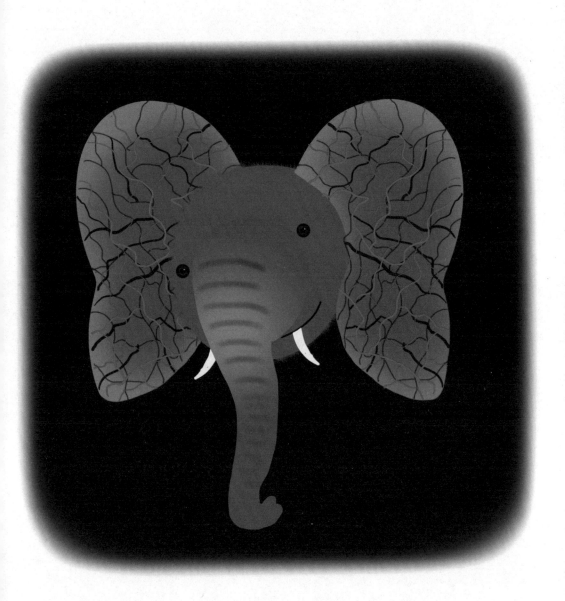

... thousands of veins and arteries ...

......在寒冷的夜晚我应该怎么做呢?

... what do I do on a cold night?

"那在寒冷的夜晚我应该怎么做呢？"小象问道，"我的耳朵这么大，那些欺负我的家伙取笑说可以当毯子用。"

　　"你的身体是很巧妙的能量调节器，在夜里它能减缓血液在你耳朵里循环的速度，所以你不会觉得太冷。"

"And what do I do on a cold night?" the calf asks. "My ears are so big, those bullies joke they could serve as a blanket."
"Your body is such a clever regulator of energy, that at night it slows down the circulation of blood to your ears, so you do not feel too cold."

"哇，太棒啦！妈妈，我现在为我的耳朵感到非常骄傲。"

"你应该如此。我们确实是世界上所有象里耳朵最大的，而大耳朵也在很好地为我们服务。"

"确实如此！"

……这仅仅是开始！……

"Wow, impressive! Mum, I am now so proud of my ears."
"And you should be. We do have the biggest ears of all elephants in the world, and they serve us well."
"They do indeed!"

... AND IT HAS ONLY JUST BEGUN! ...

......这仅仅是开始！......

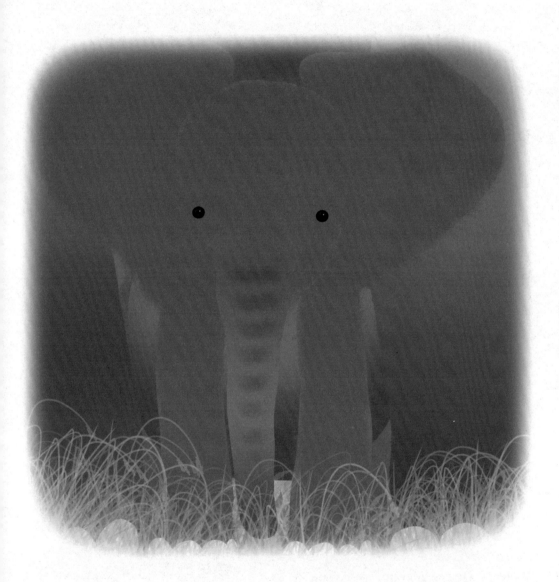

... AND IT HAS ONLY JUST BEGUN! ...

Did You Know?

你知道吗?

在已知的校园欺凌事件中最常见的起因是相貌、体型和种族，所占比例分别为55%、37%和16%。

The reason for being bullied reported most often by students in schools are: looks (55%), body shape (37%), and race (16%).

参与欺凌的学生在以后的生活中出现学术问题、药物滥用和暴力行为的风险增大。

Students who engage in bullying are at increased risk of academic problems, substance abuse, and violent behaviour later in life.

Students who experience bullying are at increased risk of poor school performance, sleep difficulties, anxiety, and depression.

遭受欺凌的学生更容易出现学习成绩差、睡眠困难、焦虑和抑郁的现象。

While humans use ears to hear, elephants use theirs to also regulate body temperature. While elephants in hot climates have big ears, the mammoth in the cold arctic had smaller ears as they needed to retain body heat.

大象跟人类一样能用耳朵来倾听，还能用耳朵来调节身体温度。生活在炎热地区的大象拥有巨大的耳朵，而生活在寒冷北极圈的猛犸象因为需要保持身体的热量，耳朵就小得多啦。

Warm blood travels to the elephant's ears and cools off when exposed to wind or water, circulating cooler blood back into the body.

当大象身体里温暖的血液流经耳朵，接触到风或水的时候会变凉，温度下降了的血液再循环回到大象体内。

Beauty cannot be judged objectively – what one person finds beautiful or admirable may not appeal to another.

美是主观的。一个人觉得美丽或值得称赞的事物可能对另一个人而言毫无吸引力。

The jackrabbit is an American desert hare. They are born with full fur, with eyes wide open and are mobile within minutes of birth.

长耳大野兔是一种美洲沙漠野兔。它们生来就带有一身的皮毛，睁着大眼睛，出生数分钟内就可以活动。

The fennec fox has the biggest ears of any animal in proportion to its body size. Its ears can be half the length of body. This allows the fox to hear prey moving around underground, and keeps it cool in desert conditions.

按耳朵占身体的比例来说，耳廓狐是所有动物中耳朵最大的。它们的耳朵可以占到身体长度的一半，使得这种狐狸能够听到猎物在地底下移动的声音，并让它们在沙漠中保持凉爽。

Do you agree with everything your friends consider beautiful?

你与你的朋友对美的看法是否完全一致?

Are ears only there to hear?

耳朵只是用来聆听的吗?

Would you poke fun at someone who has bigger ears than average?

你会嘲笑那些耳朵长得比一般人大的人吗?

What is most important for survival when you are in the desert: to cool your body down or conserve water?

当你身处沙漠的时候,要想生存最重要的是什么:给身体降温还是保存水分?

Do It Yourself!

自己动手！

Take a moment for self-reflection. What are your strengths, and what are your weaknesses? What are the strengths you have that no one knows about? And what are your weaknesses, that everyone seems to know about? Identify something about yourself that would surprise others when you share it with them. Ask them to also share something about themselves and you may develop new friendships based on the discovery of each other's strengths.

花一点时间自我反思。你的优点和缺点分别是什么？你所拥有的不为人知的优点是什么？而可能人人都知道的缺点又是什么？找出一些关于你自己的事，在与别人分享时，确认这些特质是否会让他们感到惊讶。让别人也分享一些他们的故事。基于这种对彼此优点的探索，你们可能会发展出新的友谊。

27

学科知识

Academic Knowledge

生物学	居住在沙漠的动物需要保存水分，它们不能通过流汗来降温，大耳朵能帮助它们调节身体的温度；皮肤的不同作用。
化 学	辣椒能增加血液流量，清除动脉阻塞；空调使用化学制冷剂，将热量从屋内传递到屋外；大象散发热量不需要化学制冷剂。
物 理	耳朵外侧的血管在舒张过程中，将温暖的血液流通到耳朵，热量在此处散失到周围较冷的空气里；在干旱的栖息地，耳朵降温是一种水分保存机制，可以减少通过喘气或汗液蒸发降温。
工程学	人类制造出了需要用压缩机、冷凝器、蒸发设备并消耗大量能源的制冷系统，然而大象自身的降温机制完全不需要这些。
经济学	由于城市热岛现象，给建筑物降温的花费不断攀升；热岛指的是热空气从建筑物中排出，在周边制造了更热的大气层，从而导致保持建筑物内部温度凉爽舒适的成本更高。
伦理学	我们怎么能在欺凌事件发生时袖手旁观呢？避免以暴制暴；需要找到正确的方法，引导欺凌者将精力投入到积极的行动里。
历 史	"Beauty is in the eye of the beholder." 这一说法最早出现在公元前3世纪的希腊文学作品中，后被莎士比亚用在了《爱的徒劳》中。
地 理	美国50个州均已实现反欺凌问题的立法化，规定学校中的欺凌事件是违法的。
数 学	一冷吨空调制冷量大概能满足1 000平方英尺面积的制冷。这一数字没有考虑墙和窗户的方向、建筑面积相同时单层和两层房表面面积的区别、保温层和不同建筑之间空气流通的区别、居住者数量以及其他因素。
生活方式	欺凌事件的数量在上升，在一些国家和某些环境下已达到泛滥的程度。
社会学	欺凌事件跟冲突事件是不一样的，冲突中的当事人被认为是平等的，而欺凌往往给人社会力量和身体力量的不平衡感；欺凌有4种类型：情感上的、言语上的、身体上的以及基于网络途径的；法律欺凌指通过提出琐碎、重复或繁重的诉讼来威胁被告服从当事人的要求，这一情况的发生不是因为当事人的诉求有多少法律依据，而主要是由于被告无力承担法律诉讼费。
心理学	欺凌的动机可能是嫉妒或怨恨；欺凌往往被用作一种隐藏羞耻、焦虑感或增强自尊心的工具；对于儿童及青少年来说，被欺凌或成为欺凌者的主要风险因素是缺乏解决社交问题的能力；被欺凌的儿童害怕去学校，抱怨头痛或缺乏食欲，对参加学校活动以及陪伴朋友、家人失去兴趣，还会常常感到悲伤。
系统论	社会中的欺凌影响整个社会的组织，并且对孩子的未来造成影响。这种影响不只是孩子们精神上和身体上的健康，还包括对他们未来财富和生活质量的影响。

情感智慧
Emotional Intelligence

小象

小象对自己因为大耳朵而被取笑一事有很大的压力。她为自己感到羞耻，并且不喜欢自己的外表。她不相信她的父母结婚的原因是爸爸认为妈妈漂亮，而觉得是因为她爸爸没有其他的选择。小象不知道耳朵在调节身体温度方面的作用。她提到欺凌者取笑她的耳朵是用来遮阳的，这也再次确认了欺凌的发生。小象喜欢在水里玩耍，这可以给身体降温。她发现自己耳朵部位的皮肤很薄。她明白了耳朵可以帮助身体降温。小象回忆欺凌者们说她的耳朵大到可以当毯子用。当她明白并且完全了解了自己耳朵所有的功能以后，小象对拥有大耳朵感到非常自豪和满足。

象妈妈

象妈妈对于孩子被欺负感到很担心，但并没有建议报复或是采取攻击性行为，而是先指出美取决于个体的欣赏水平。她尝试让小象发现自己的优点。象妈妈问小象除了聆听，耳朵还有什么其他功能，让小象自己发现关于耳朵的知识。象妈妈帮小象明白了耳朵有助于在白天和夜晚调节身体的温度。她在解释这种自我调节能力的同时，成功地把小象的压力转化成了作为族群一员的骄傲。

艺术
The Arts

拿出纸和铅笔，分别画出朋友的、兄弟姐妹的、爸爸妈妈的以及一些演员或电影明星的耳朵。你注意到了什么？每只耳朵都有其独特的样子，有些耳朵跟其他耳朵的差别非常大。莫扎特的耳朵与众不同，这也为他为什么是一个如此有天赋的音乐家提供了一种解释。当你画了6只甚至更多不同的耳朵以后，让你的朋友们认一认其中哪些是属于你的家庭成员的。他们能很快分辨出来吗？

思维拓展
Systems: Making the Connections

欺凌现象可能总是存在。无论如何，处于网络时代，我们现在更清醒地认识到身体上和心理上的威胁。网络欺凌可以是匿名的，包括威胁、散播谎言以及将他人真实或感觉上的缺点暴露在世人面前。

研究证明，当班级人数少，老师与学生的互动比较直接和个人化的时候，欺凌几乎不会发生。欺凌事件通常出现在从小学到初中直至高中的过程中，尤其是学校合并班级的时候。欺凌事件的受害者承担着贯穿他们一生的心理、身体和经济上的伤害，一些人可能应付不了，甚至自杀。研究显示欺凌者通常在其他学生甚至老师中很受欢迎，并被认为是群体里最酷的成员，而受害者则完全相反。

欺凌者手握大权，具有力量并且自我膨胀。受害者往往没什么朋友，而且不知道如何制止欺凌的发生。更糟糕的是，受害者通常相信他们自己是应该被责备的，他们本来就有错，所以发生在他们身上的欺凌是他们应得的。这变成了一种自我实现的预言。这些弱势的孩子变得更加沉默寡言、唯唯诺诺，变得更加脆弱。

欺凌不是少数个体的问题，而是一种社会层面的问题，由于欺凌者和受害者是过渡型的角色（而非个人的永久性特征），因此亟须改变这种社会现象。改变的关键是让欺凌者发挥特殊的作用，将他们的精力转移到积极作用上。之后，要引导受害者走出困境，扭转他们的看法，让他们知道自己有机会发展新的友谊。

动手能力
Capacity to Implement

你曾经目睹过欺凌事件吗？跟你的朋友和老师们讨论一下，找出一些欺凌者可能用来攻击受害者的理由。讨论这些理由，看看它们是否属实。如果这是受害者的缺陷，你能如何帮助受害者发现自身的优点，并基于这个优点来鼓励受害者？教室里的领导者（被认为是最酷的人）能不能发挥作用，将精力转移到对学校、同学有益的事情上？

故事灵感来自

This Fable Is Inspired by

娅娜·尤沃宁

Jaana Juvonen

娅娜·尤沃宁出生于芬兰的一个小镇，拥有芬兰图尔库大学教育心理学硕士学位。她现在是美国加利福尼亚大学洛杉矶分校的一名发展心理学教授。她进行了数十年的研究，验证欺凌者有高度的自尊心，并且经常被他们的同学甚至老师看作学校里最酷的那类孩子。她发现欺凌通常发生在中学而非小学。她在如何创造一个有助于消除欺凌现象的环境方面给学校提出了许多建议。

31

图书在版编目（CIP）数据

冈特生态童书.第四辑：修订版：全36册：汉英对照 /
（比）冈特·鲍利著；（哥伦）凯瑟琳娜·巴赫绘；
何家振等译.—上海：上海远东出版社，2023
书名原文：Gunter's Fables
ISBN 978-7-5476-1931-5

Ⅰ.①冈… Ⅱ.①冈… ②凯… ③何… Ⅲ.①生态环
境-环境保护-儿童读物—汉、英 Ⅳ.①X171.1-49

中国国家版本馆CIP数据核字（2023）第120983号
著作权合同登记号图字09-2023-0612号

策　　划　张　蓉
责任编辑　曹　茜
封面设计　魏　来　李　廉

冈特生态童书
大耳朵
[比]冈特·鲍利　著
[哥伦]凯瑟琳娜·巴赫　绘

贾龙慧子　译

记得要和身边的小朋友分享环保知识哦！
八喜冰淇淋祝你成为环保小使者！